Magnetic North

Sea Voyage to Svalbard

JENNA BUTLER

Published by

The University of Alberta Press
Ring House 2
Edmonton, Alberta, Canada T6G 2E1
www.uap.ualberta.ca

LIBRARY AND ARCHIVES CANADA
CATALOGUING IN PUBLICATION

Butler, Jenna, 1980–, author
Magnetic north : sea voyage to Svalbard / Jenna Butler.

(Wayfarer)
Issued in print and electronic formats.
ISBN 978-1-77212-382-1 (softcover).—
ISBN 978-1-77212-421-7 (EPUB).—
ISBN 978-1-77212-422-4 (Kindle).—
ISBN 978-1-77212-423-1 (PDF)

1. Butler, Jenna, 1980– —Travel—Norway—Svalbard. 2. Ocean travel. 3. Climatic changes. 4. Feminism. 5. Svalbard (Norway)—Description and travel. I. Title. II. Series: Wayfarer (Edmonton, Alta.)

G778.B88 2018 919.8'1045
C2018-902559-X
C2018-902560-3

First edition, first printing, 2018.
First printed and bound in Canada by Houghton Boston Printers, Saskatoon, Saskatchewan.
Copyediting and proofreading by Peter Midgley.
Map by Wendy Johnson.
All photographs by Jenna Butler.

The University of Alberta Press is committed to protecting our natural environment. As part of our efforts, this book is printed on Enviro Paper: it contains 100% post-consumer recycled fibres and is acid- and chlorine-free.

The University of Alberta Press gratefully acknowledges the support received for its publishing program from the Government of Canada, the Canada Council for the Arts, and the Government of Alberta through the Alberta Media Fund.

Canada

Canada Council for the Arts
Conseil des Arts du Canada

For Andrew Miller, Dana Sachs, Rosalind Hudis, Susan K. Salzer, and Todd Wronski, with heartfelt thanks for our Hawthornden spring.

& for Thomas Lock, true north,
who draws me home.

The wild places that we love to death are cowed by human tumult by day,
but return to their true nature after sundown.
Darkness has become our last wilderness.

—SID MARTY

Contents

xi The Journey

1 Lines Toward Ice
7 Pyramiden
13 Ornithomancy
19 Night
23 Bone
29 The Men at the Edge of the World
35 She Becomes the Ocean
41 Arctic by Air
47 Afloat
53 Barentsburg
59 Cusp
65 Postcard from Svalbard
71 At the Face
77 Threads
83 Leaving Days
89 Song to the Boreal

99 *Notes*
101 *Acknowledgements*

The Journey

IN 2014, I held a position as writer in residence onboard an ice-class barquentine sailing vessel in the Norwegian Arctic. For two weeks in mid June, over the summer solstice, my colleagues and I sailed around the islands of Svalbard, investigating Spitsbergen's bays and fjords, its mining towns and whaling stations. During that time, we learned a little about the routines and needs of a masted schooner, and we learned a great deal about the environmental degradation taking place in the high Arctic because of climate change. From hundreds of feet away, we observed glaciers calving through the long, bright nights of the Arctic summer. We photographed locations where the glaciers had completely receded from the rock face for the first time in recorded history. And we visited the ruins of whaling stations, the current mining town of Barentsburg, and the ghost town of Pyramiden, curious about the ways in which humans had coaxed a living from the rocky bays and inlets of the islands. Everywhere we looked, we saw human intervention etched onto the land in the form of mine ruins, processing effluent, and piles of whalebone, but we also saw the tight-knit, multi-ethnic town of Longyearbyen blinking its lights over the harbour. Community strove to balance industry.

The Arctic taught us a great deal over the space of a few brief weeks. We learned what had drawn the first polar explorers to the island and what continues to entice people to the modern settlement at the top of the world. Most of all, we came to understand more clearly what had led us to pursue artistic work in such a remote place, what called us to the uncanny Arctic landscape, and that we would all carry a piece of Svalbard back home.

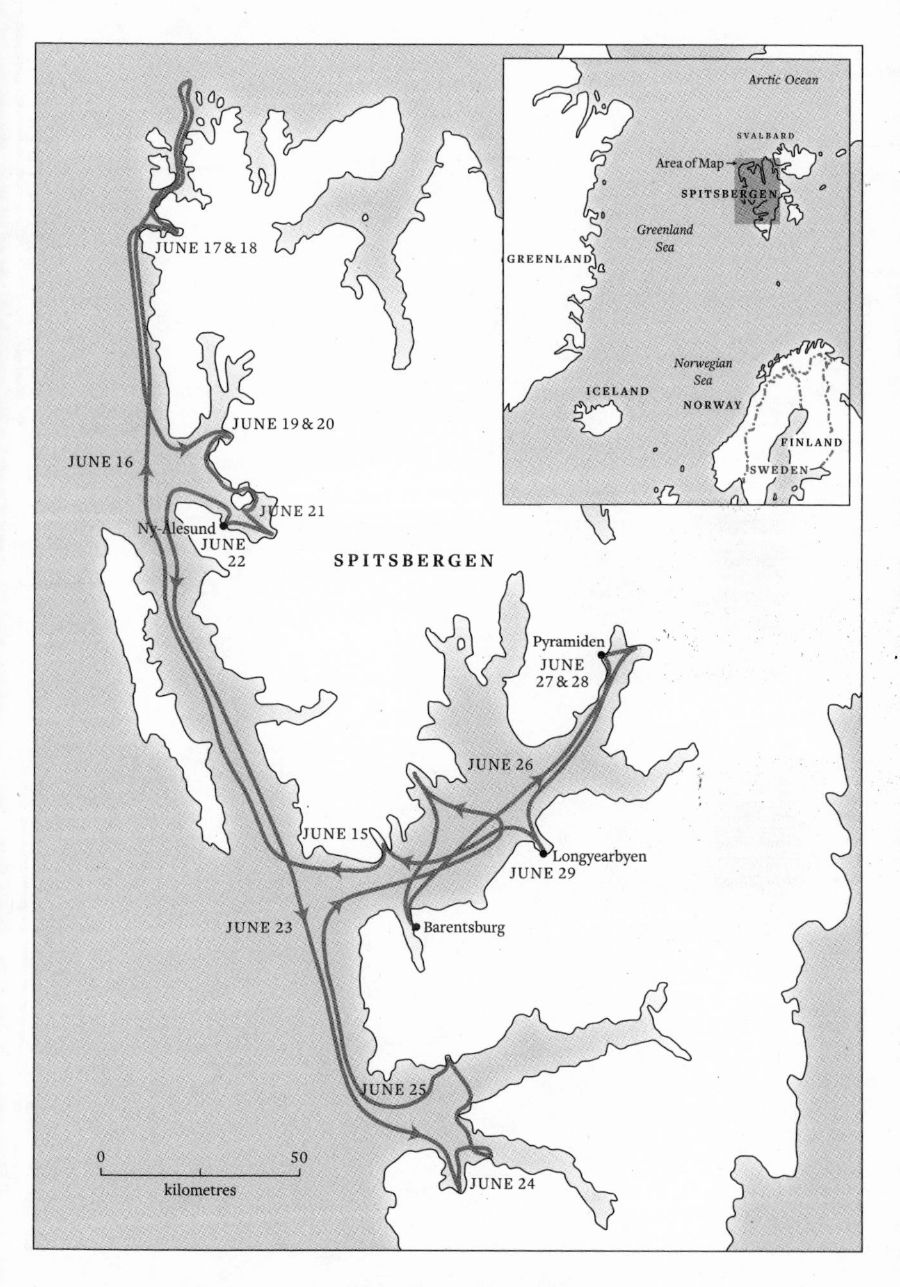

Voyage of the Antigua, *Spitsbergen, Svalbard, June* 2014.

Lines Toward Ice

the whole pageantry

of the year was
awake tingling
near

the edge of the sea

—WILLIAM CARLOS WILLIAMS

1

Further north than I have ever been, hills emerge from the sea, obdurate, popping knuckled seams in the day's meagre heat. Sun on snow, tideline clotted with scrims of pack ice.

There are no animals here but rock itself gone swayback up from the water. My palms tingle in my pockets, singularly attuned to prairie and its capaciousness, a rigour of skyline. These hills beckon some obscure phrenology, a laying-on of hands the only way to suss out the true form of a country obscured the majority of its life by quick-moving drifts. Even now, a shifting of the wind, snow winnowed out across the rocks and just as suddenly eddied in on itself, collected again. Obfuscate and make plain, sleight of hand, the calculated reveal. A good poker face.

2

What I know is the knife-edge of boreal forest, gantry of muskeg spruce
hoisting ravens against the clouds. My eyes attuned to periphery,
horizon line muggy with deer flies, nascent thunderheads, prairie fleece.

This land is pleached square against the skyline, unforested entirely.
Over port, the gannetries rise in a consortium of granite. Birds in their
millions skirl and banter, sound so all-encompassing that it ceases to be
sound, becomes substance. Livewire spark, electric. The cliffs coruscate
in birdlime; at midday, they are impossible to look upon. A different
kind of snowblindness, those scabbed rocks, the sound they house.

3

We nose into still water leeward of the rocks, the kittiwakes haranguing, mewling low on bold wings. Shunt aside the growlers clanging against the hull, put ashore at the only strip of beach visible below the ice crust. Boreal navigation I know: wading through Labrador tea plants jittering with early bees, plush moss, the unfurled heads of ferns. On the beach, moss campion fingers itself over the south side of the rocks, blossoms frantically and in miniature.

Here we expect pebbles, find instead a midden, narwhal and harp seal, gannet wingbones thinned to needles. A landscape where one goes slowly, eyes on the ground, a thousand years of human habitation rustling under our boots. Kneel to recover the lost feathers of Brent geese, gone now from the beach, out over the water to Taymyr.

4

At home, winter's scent is quantifiable: black spruce and frost, bistre of
tamarack roiling in the wood stove. Out here, the air has been scrubbed
clean. Bergs heft their indigo bones like naves in the sunlight. We look,
as we have always done, to sight something looking back at us, but there
is only the expanse of midden beach and its disarticulated moonscape.
Nothing here precisely for us or against, just a great unembodied
presence. Narwhal horns cross their shadows from the abandoned
heaps, unperturbed as ice, or time.

Pyramiden

We have become experts in analysing what nature can do *for* us, but lack a language to evoke what it can do to us. The former is important; the latter is vital.

—ROBERT MACFARLANE

1

Sasha is the lone figure on the wharf.

Momentarily, we are time-blinded, thinking Czarist Russia. He greets us in felted black *cherkesska*, polar bear rifle angled along his back: Sasha, improbable tenant, unconvincing miner.

What they do in this town weeping its coal ichor into the sea: harvest metal. Fourteen of them, a pack headed by Sasha, self-proclaimed tour guide. They come from Russia because there is no work; here, there is no work, but at least they can pick metal, pack it by the shipping container for the resupply vessels from Murmansk. The law leaves them to their self-imposed isolation, these men without homes, without countries.

2

Every building abandoned to gulls and gannets, a squalling, shrieking mess. Stink of guano so heavy it smothers, paints the tongue. Even the hospital has gone to birds, though Sasha explains that the plants have also taken over. Their wild, wastrel tendrils pushing through the outside vents, pressing palely against the glass. Sasha himself has not fallen ill since leaving Russia a decade ago, but others have raided the hospital for tablets and syringes. Morphine for the bad days. Sasha turns his collar, refuses to enter. Says the air is poisoned with mercury from broken thermometers after the town was left. *Was left.* Curiously passive, a child, a family walked away from, abandoned. But here, the whole town left, more than a thousand in clots over years.

We turn our backs on the hospital compound with trepidation, uneasy with the blank windows behind us. The greening glass, leaves spreading their veins like suicide or palmistry.

3

One room remains, half Russian teashop, half bordello. Napped crimson wallpaper and gilt mirrors resurrected from birdlime, parquet floor high-glossed with lemon polish. We drink Kubanskaya, all the fugitives will touch; *original*, they explain. Kristall and Moscow-Petushki; we drink to times and places we do not understand, the bitter, amputated stories of these men.

They were meant to be miners. After the second war, Russia remembered Pyramiden, abandoned against the Nazis. Clapped men in ironclad contracts, sent them back across the ocean into the shankbones of the mines. But the hills bled their bones out in the 40s; left to their own devices, rain and snow took the coal out to sea, topped the shafts with inky cesspools. And so the men pick metal, the cuts on their hands rimmed with black, coal inching toward their bones. Drink vodka in the evenings, in the one room pushing back against a settlement's decay, Russia boiling out of their blood as the coal works in. For six months during his first winter, Sasha tells us, he did not leave the common room. It takes time, he says, to armour yourself against that much dark.

4

The one surviving telephone is an iron ramshorn, Cold War–era megalith on the way to the docks. It works in one direction only: outward, to Russia. Open channel in a diminished conversation between the men who have chosen this life and all they have left behind. The mine spilling its framework downslope as groundwater chases panels free, Pyramiden bleeding its story each runoff clear to the Arctic Circle.

Ornithomancy

In September, countless sand and house-martins jazz above the river, taking insects from the surface, from the air, thousands of birds kissing the river farewell. They creak, a sound like the air rubbing against itself. Summer is everything they know; they're preparing themselves, sensing in the shortening days a door they must dash through before it shuts.

—KATHLEEN JAMIE

1

Pleated bone flutes open against rock. Arctic tern caught sunning, breast jimmied open by the rummagings of the fox. Little left but the halfmoon curve of her beak, a startlement of pinions.

The sun is directly overhead twenty-four hours a day. Death leaps at me with the intensity of searchlights. Bones bleach and fissure over a matter of weeks, the only shade directly beneath, where the blood slips, pools, sprouts a small tuft of grass almost overnight. Death is immediate, and life, in this pressure chamber of northern summer, is nearly as quick.

The terns nest with a sort of desperation, clenched in their thousands amongst the stones, splayed and vulnerable. My walk a drunken weave, skirting that nest and this, the airborne ire of mothers. Livewire buzzing; the young that do hatch are raised in a ratcheting of *fear, fear, fear,* haptic filaments needling through the spaces in their bones.

2

Off Spitsbergen, Fanshawe lifts its dolerite skirts ribbed with whalebone and eggshell. Guillemots rasp into the air in their thousands, adamantine hammer of sound that drills in somewhere behind the eyes. I feel the birds as much as see them; in my blood is a new element, a battering of wings, mad clangour.

The land is contoured in a way that breeds sound: it sloughs from the bird cliffs, heaves over the water in a rich, acrid roll. Along the shoreline, sound clatters between the growlers, barks chips off the larger bergs. All through the afternoon, champagne tinkle of ice chips as slabs greet each other across narrow channels in the green.

Windburned, eyes closed, this: beneath the keening of bergs, a deeper thresh of glaciers calving, creaking with sun. Sound of earth, her bones, wide russet bowl of hips splaying open. From these sere flanks, her dessicating body, what a sea change is born.

3

Magdalenafjorden at the solstice, scree slopes greening overnight as little auks come in from the sea, fast and fetid life clamping itself to the stones. Blue eggs erupt from the cliffside like a party trick, their shadows sewn beneath them.

Imagine what it is to be called through the bones, daylight playing itself out in your marrow. This knowing that draws you in from hundreds of miles out to sea, slices you cleanly through the spindrift to shore. What it must be to understand the pull in your blood this way—
an old, old calling.

What it must be like to know your direction without fear.

4

Eggs that mysteriously appeared and vanished again, palmed, feinted, perfect diversion, child's birthday trick. When your nieces hunted up your sleeves, small hands darting like voles, how the laughter would roll out of you while the secrets stayed in, stayed put. Somewhere was a prize and you weren't telling; better than a chocolate shell, something real. You imagined it held delicately against your teeth, your tongue, certain that when they pried your lips apart, the secret would scattershot into the world like a small bird.

You held my hips that way, as though you were framing sky, permitting egress, flight. As though my body, too, could be palmed, appeared and disappeared like a vaudeville trick; somewhere a small skittering of claws, tattoo of wings. What we didn't know, then: how winter breaks down a nest from the outside in, weathering its bindings to silver. The way the cage of the body can break open at a touch and nothing flies forth. No sound, no tumult of raddled feathers. Just empty air, and light.

Night

Declension
stalks the snow cover.

It wonders how many people it is
how many mouths.

—MONTY REID

1

There's no place to hide.

Night is non-existent, circadian cranked to overtime. No twilight, no dusk, no dawn. Same abbreviated shadows on the road. We jitter, grateful for anything that dims the light: snow squalls, cloud banks, duty-free Svalbard whiskey. Something in us paces, sleepless, sits on its haunches and observes morning rituals of black coffee and *gjetost*, uneasy company.

Svalbard bleaches its bones four months of the year, stretched thin by sun like a drum on a frame. Its wild places crack in the heat, tundra canvassed out. Pinned, strained.

Twenty-four hours a day, the hunt. Reindeer on the campion flats, polar bears along the beaches. Fox dens green to uncanny mounds, small death forests visible from the air. So much sun that entrails, clapped blue against the rock, morph in an instant to chickweed, pygmy buttercup, Svalbard poppy.

2

Blood runs amok, sunspurred, no easy resting place. Birds at all hours, a ferment of terns over stone-couched nests, fulmars ricocheting from the cliffs. Inner harbour, midevening, fourteen belugas breaching, heat calling them to the surface like bergs on a slow turn.

Darkness becomes mythic, vein-shot, red light through pinched eyelids. We are bitter, amped. Our bodies want to move, to work, to love; we burn out in three-day stints, sleep through afternoons, emerge into perpetual midday. Tremors in our hands, daylight bubbling in bloodstreams.

What we took for granted: the relief of the dark. How concealment, too, is a sort of blessing.

3

Aboard ship, dark becomes an interior country.

Thirty people in six hundred square feet, the middle of the Greenland Sea, the top of the world.

We learn what the explorers knew: that darkness is not a right but an indulgence. You do what you need to find it: the bar is open as long as the sun is up. Love, too, is a kind of darkness, a momentary way of abandoning the body. Forget where you left yourself. Don't look for tracks.

The wilderness is in the darkness and the darkness is internal, layered over, contents under pressure. Coalescing into something much less human, more keen. If darkness is the fringeland, sidestep the mirror. All the world in your eyes staring back.

Bone

Language has gotten restless, it's true
but that doesn't mean it wants you to stop
pulling at its edges.

—MONTY REID

1

When the snow recedes, slinks itself upslope toward glacial till, Svalbard manifests a landscape of bone. Old outposts bleach their siding under constant sun; in sheltered bays, whaling stations lean inland, stunted by polar wind.

The midden heaps are visible from sea.

We ship sail, canvas luffing, and anchor offshore. What seemed ragged snowpack exposes itself on the beachhead, becomes a bone bed splintering under Arctic noon. Hundreds of belugas strewn like a jigsaw, heaped and abandoned. As though a bowerbird had gone amongst the stones, taking trophies, sowing vertebrae with moss.

2

Bone frays as it ages, as wood does. Goes porous, fibrous, greens and rots. Every midden heap a garden.

In some places, skulls have been arranged by others, visitors or whalers. They take communion in half-moons, nightmare fairy rings crescented along the beach. *Muscovy*, *Noordsche*; the evolution of the whaling companies shifts under your feet if you dig. Lift the skulls from their nests of campion and more bones gleam beneath. The middens are built into beachheads, sand and stones packed like ice around a wound. Hundreds of years of trade crack into the shingle.

3

From *Antigua*'s decks, I feel a kinship in my body. Looking out over the beachheads firing white in the sun, pods upon pods of bowheads, belugas driven up on the shore and flensed, it is impossible not to feel a peculiar resonance. Bone recognizes its own in all places; it calls and calls, firecrackering through the marrow.

What can any of us hope for? A pair of hands lifting us from the rubble of our end, aligning us in such a way that all of our darknesses break open and gleam.

4

A woman in this place feels its sudden excesses. Tern eggs tumbled in the beach stones, Main Street straddling a flume of ice sheet runoff. The palatial blue crags of glaciers and bergs.

Things here have a propensity to go belly-up when least expected. Off Edgeøya, the bergs flip their lids on hot days like grease dancing in a smoking pan; *pop*, and the ground lifts, reforms beneath my feet. How like life—as I watch a ringed seal barely intuit the shift, heave itself clear in alarm when the berg bottoms over. Just as I've balanced my weight, taken the ground's measure, everything teakettles.

It's the beauty of this place that catches me off guard. The light up from Port Longyear that paints the mine ruins a thin gold, reindeer cropping grass by the cemetery. It's the paradox of knowing that nobody can be born on Svalbard or die here—those burdens flown elsewhere. The only bones that rest here are those of the animals whose land this is, whose land this has always been.

The Men at the Edge of the World

Judge a moth by the beauty of its candle.

—RUMI

1

It is a hard land of few women.

Perhaps the Vikings turned their longboats here once. *Svalbarði*, the cold rim, the edge of the world, but there are no traces of iron men in the stones of Spitsbergen. Perhaps they were a Norwegian hope, a desire for something wilder than the land itself that could be pitched against December's always-night, June's unsetting sun.

Svalbard is a land of traces: what dies, lingers. The bone beds of the whaling stations, the outposts with their ragged timbers overlooking the straits. Each thing that lives its space on the island casts some small shadow, a sundial arm of birth and death. Only humans in this place foreshorten the clock, turn away those about to die, to be born.

2

In Nordaustland, Pomor crosses rise from the hillsides like scurvy teeth.

Four hundred years of summering on this land, of knowing the archipelago and its waters. Four hundred years of mistrust: of the winters, the cold. The long dark that scratches its bones down through glacial till as the light begins to fade.

Early spring, the drift ice jangling and clanging, drumming against the wooden boats. Imagine this new migration: from the mouth of the White Sea, Pomor men drawn out across the salt to Spitsbergen. *Murmanskoye Morye*, Sea of the Norwegians, heaving with pods of bowhead and beluga, rattling with birds.

Too many of the crosses have come down to time, turned to firewood in the deepest winters. Those left swing their pinwheel arms in the wind off the glaciers. If you look to their roots, if you count the base rings where the wood has been driven amongst the stones, you can hear the faintest breath of Karelia, scent the pine forests of Kandalaksha out along the Gulf. There is no wood on Spitsbergen, every plank brought from away. What is not pilfered for fire is permitted to stand, uncanny and apart, roots slagged in permafrost a thousand miles from home.

3

Barentsz went after China and found polar bears instead. Bjørnøya, Bear Island, rippled its grey flanks up from the sea, threw his men beasts of a size and shape beyond imagining. Humped and fleet, these bears hunted because they could, the pleasure of it disconcertingly human. Under the searchlight Arctic sun, they incandesced with the stink and joy of meat.

The Pomors grew wary of winter. Raptor winds off the glaciers, the perpetual thunder of icefall and avalanche. Perhaps Barentz saw the lingering crosses, their shadows like giants along the deepening snow. Watched sunlight caress the tips of the spars before it vanished over the edge of the world.

In the camps at Novaya Zemlya, forced wintering, his men starved and bled, limbs a welt of scurvy bruises. Perhaps Barentsz heard Holland at the end, thin and green and distant across the curve of the Norwegian Sea.

4

Downshore at Smeerenburg, blubber ovens hunker into the banks, protecting hollowed guts from the wind off the sea.

This coastline is a place of extremes, stone heated to a gashed red over the summer months, stink and smoke of bowheads flensed of fat. In winter, when the pods have moved out of reach and the ovens are silent, waiting, it is a place that cannot sustain a village through the months of dark. Smeerenburg is built solely on summer, built solely for death.

Women in this place are conspicuous in absence, though whalers tend the ovens day and night through the long Arctic summers to make lamp oil and soap. They comb the corpses for baleen to be fashioned into parasols and corset stays. This place of no women is stripped, scraped, flensed, boiled in iron; piece by piece, it is dismantled and sent back over the water. Each spring, the whaling ships go north again, dodging pack ice; each autumn, a seine of complicity skeins back across the ocean toward home.

She Becomes the Ocean

I fancied you'd return the way you said,
But I grow old and I forget your name.
(I think I made you up inside my head.)

I should have loved a thunderbird instead;
At least when spring comes they roar back again.
I shut my eyes and all the world drops dead.
(I think I made you up inside my head.)

—SYLVIA PLATH

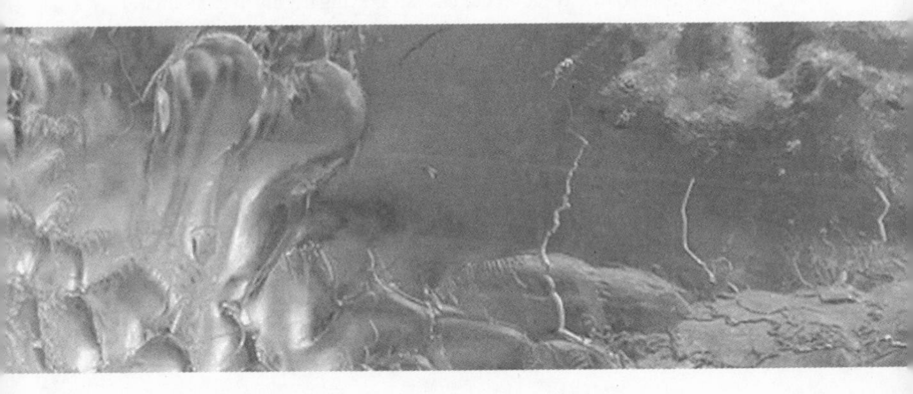

1

When he signs his name to the list, he is gone to the ends of the world. Thousands lose their lives to the drift ice, caught unawares. Silver flip of a coin and they all go under.

She freights her days against his absence. Takes in sewing, laundry, fingers riddled with needle-bites. Cheap tallow smokes up the walls, reeks of rendered fat; it blurs her eyes, dulls the fine stitches. If he returns in the autumn, they may do well. If he never comes, she will starve.

Each day a pushing back against the seasons. Arnhem greens into summer, delicious shadows of plane trees, the sweet white clover. In the back of her mind, she imagines what ice must sound like, clinking against itself on ocean swells. She has seen it in the kitchens of rich houses, slivers of ice as perfect as claws, dizzying themselves against the glass in pitchers of cassis.

2

The Sint-Jansbeek purls down from Zijpendaal. This is the water she knows, clear on the good days, fouled much of the time with offal, tannery runoff. Last month, they drew a girl from the water, slow dance of her body face-down in an eddy. Skin loosened, bated by current and streaked with hemlock from the curing vats.

She wants to know water, each mile of it that separates them. She has never been to the sea, cannot imagine water so salty that it burns instead of quenches, leaves a thin rime where it falls. Over the months, water works into her dreams: the open sea, the ice. A looseness to her marrow, something moving, subterrain. Water welling up from deep places, the permafrost addling to sponge.

She wonders whether the wild lick of the sea keeps him from home. The taste of salt a kind of forgetting.

3

The ships make port on a fine day in September, heeling in before the autumn clippers. Their holds are full of tusk and bone, lamp oil smoked down on the shores of Jan Mayen, Spitsbergen baleen. He brings her two quarts of oil to replace the tallow candles greasing their stink up the walls, sparks Arctic midsummer in her sewing garret. An astonishment of light.

He brings her a span of baleen for the corset she could never afford to wear, bleached fine and white through the heat of June. The wands of baleen slip through the corset's plackets like the masts behind the rigging, spanning a fine form. Last, she draws from the sailcloth packing a busk the height of her torso, whalebone the colour of tea-dimmed china. Along its length, he has scrimshandered the sea: a rope of guillemots, a whale pod breaching. The busk keys into the corset's centre rib and she is laced into the Arctic summer. Blood pounding in her ears might be the waves soughing. She settles the stays, lights the lamp. Licks salt from each finger.

4

He is gone in the spring with Barentsz' crew.

The lamp is empty, magic oil doled out through an Arnhem winter. She lays in tallow and lengths of green reed, her walk a slow sway, the shoals in her belly. The corset languishes on its hook behind the door, baleen crackling softly. The stays feather empty as pinions.

She is waiting for another gift of light, the Arctic brought home to her in scabby glass. The whale's body ensconces the house, its empty rooms. The whale's likeness on the busk she has drawn from her corset like a blade.

She senses the winds returning: no passel of sails down the coast, the pack ice closing ranks in the archipelago. When the news comes, he is half a year dead.

Their child will be born in ropy tallow light, blood and meat and earth. He will carry Svalbard in the tint of his eyes, the look she will catch on him, slow turn of deep-water ice. Something coming up from beneath. Even the taste of salt eventually fades, inland. The tides now in the bowl of her hips, their irresistible splay.

She slips the busk between her teeth and bears down.

Arctic by Air

Historical Note: *In 1928, following a successful maiden voyage in 1926, General Umberto Nobile piloted the airship Italia from its base in Milan to the North Pole. On the return voyage, the airship crashed spectacularly near Svalbard, killing six crew members and stranding Nobile and his remaining men out on the ice. In an uncanny trick of fate, enough supplies also plummeted to the ice to keep the crew alive, in addition to a tent for shelter, and a radio. After numerous attempts, the crew established contact with the outside world, and a multi-nation rescue effort successfully brought the men in off the ice several weeks later. The rescue had its own costs, however. Famed polar explorer Roald Amundsen vanished in his plane while attempting to assist Nobile, and no trace of his end has ever been discovered.*

1

After the Great War, the Arctic reads like a who's who of aviation daredevils. Schmidt and Vodopyanov. Eckner. Amundsen and Ellsworth. Wilkins and Eielson. Nobile. Men who have seen the war dismantle their countries, pilfer childhood friends and family. Death on such a scale is seismic: it alters interior landscapes, makes men long for the simplicity of ice, the wide, clean miles between themselves and the trenches.

2

The airship is doubly blessed, and even that won't save her.

At the Vatican, Pope Pius XI speaks with the crew, offers Nobile a wooden cross to install at the North Pole. Later, the men crush a bottle discreetly against *Italia*'s flanks, offering to protect her with their own hands. She will call in all these favours, will find them wanting.

It's 1928, it's Italy, Milan jitterbugging like a flapper between the first War and the next, not yet under the Depression's heel. These are young men and the Arctic is inviolate in a globe becoming too quickly known. Nobile, Malmgren: they are hungry for frontiers, for the blue cape of glaciers spread across the top of the world.

Nobile, small and feral, his focus always the middle distance, ten steps ahead. But there is play in him: how he indulges Malmgren's request to take the *Italia* out over his mother's village, to descend to one hundred feet, blocking out the sun, low enough that Malmgren can drop a note into her back garden. When all goes awry weeks later, these are the words she will return to, incongruous from the sky.

3

In Kings Bay, the mooring mast of the *Norge* and the *Italia* still stands almost a century later. I slog through melting permafrost in the footsteps of the polar bear guard, modern Viking with her high-powered rifle slung over her shoulder, eyes scanning the hills for a quick-moving flash of white against ragged June snow.

The mast hums and clangs in the wind off the ice. The sky wheels with birds; in my mind's eye, I try to reconcile the *Italia* with this place, her incredible bulk on a slow turn from the mooring ropes, the chatter of Nobile's men echoing down from above.

This is what the Arctic brings you when you think it conquered: a sudden seething of cloud, a rime of ice along the dirigible's envelope. When you think you've mapped its boundaries, the north sends you reeling out of the sky with frost on your wings. The only thing that saved Nobile was a cosmic trick: Icarus crash-landed at the top of the world with a tent, a crate of rations, and a radio.

4

Where were the women in all this?

Perhaps it's simpler to turn north when you are not the anchor, the mooring point. *Italia*'s mast keens in the Svalbard wind, its guywires humming into this century. Think of all the masts and moorings, from Russia to North America, a hundred thousand anchor lines spanning from this cap of ice back to homes, to families. What you leave when you set your eyes on the polar north, when the wind off Larsbreen catches in your chest like some sort of witching. Those long, white, thrumming miles.

Where are the women? They are raising the children and tending the dead, navigating each morning around the sheen of ice on the floor, the vast bulk of Spitsbergen hovering like a hallucination. They are the ones for whom the north remains mythic and terrible. How it pries into their men and calls, calls. An altered interior country, even if the men come home again.

Afloat

Do not waver
Into language. Do not waver in it.

—SEAMUS HEANEY

1

I've been around boats all my life, the swallow glide of yachts, the quiet slip of canoes through mountain water. *Antigua* raises sail, though, and I am lost, seasick, nothing lovely about it. For a week and a half, my body keens into its bones before entering into a grudging treaty with the Greenland Sea.

On the prairies, you hear rain coming half a day away, see it in the blunt anvil heads of stormclouds, their long-distance rumble. Beyond June rains, water is a bounded element: frogsong in spring ditches, the duckweed curtain of prairie sloughs.

Aboard *Antigua*, I am educated in the ways of the ocean, its true face. Hypnotic rush of waves against portholes, our cabins breaching the waterline. Shipping canvas as the winds pick up, *Antigua* skittering for shore under motor, spray backhanding her decks. A consciousness of water: the Arctic grows in me this new sense, this way of hearing. The back of my mind awash.

2

I wake into thrumming noise, the porthole cover clanging, Maggie tangled in blankets in the top berth. Our cabin a welter of night breath and damp wool.

Antigua is heaving on her haunches, bucking waves hard enough to lift my feet from the flat of the bunk, then my head. The back of my skull clapping the partition wall like a gong thumps me from precarious sleep.

Seasickness reframes me, my limbs like water, my body wraithing with hunger. I clang shins on the steep pitch of stairs, aiming for the aft deck and uncanny brightness of midnight. The sea seems closer than before, gunmetal surface a raft of spume. The door throws my weight back at me: the deck has gone under, vanishing and reforming as water breaks across its length.

The prairies a thousand miles away, I feel my body gone glass, emptying and refilling with Arctic swell. Darkness and safety a trick of the mind, as distant as the long, light fields of home.

3

Morning after a night of hard sail, long hours lurching unceremoniously from the bunk as *Antigua* flees toward shore. Zodiac goes in to the beachhead, polar bear guards already triangulated to keep a gaggle of humans safe from the wild.

I am alone aboard ship, forming an uneasy truce with *Antigua*, sonorous gong of drift ice along her hull after a hell-winded night of racing. One place on her aft deck where the wind doesn't strike: here, I curl with tea and downfill, clean scent of the archipelago's water, a scurf of murres over the ice.

The ship teaches us what the explorers knew: that solitude is as rare as darkness, as necessary to sanity. Thirty people in six hundred square feet: we seek ways to be alone, fleeing for the tide line, higher into the hills if the guards allow. I am learning how to absent myself in the midst of a crowd. Slow afternoons like these, I come to a pact with the ship, spinning like a captured gull at the end of her anchor chain. Soft droop of her wings cinched into rigging.

4

From *Antigua*, these gifts:

Day Three: a minke whale that breaches and runs beside us.

Day Nine: on the headland, one polar bear, head down and tracking.

Day Eleven: a heap of walruses, blubber puddle.

Day Twelve: a sheltered bay where we hike melting permafrost, boots filling with water that has not been water for eight thousand years.

Day Fourteen: quiet seas as we motor south from Ny-Ålesund under late-June sun, saxifrage flats exhaling honey across the shallows.

Barentsburg

Let the new faces play what tricks they will
In the old rooms; night can outbalance day,
Our shadows rove the gravel garden still.
The living seem more shadowy than they.

—WILLIAM BUTLER YEATS

1

Lenin's dour face greets us on the climb from the docks.

There are two ways into Barentsburg: by tourist cruiser from Longyearbyen, or by airliner from the Ukraine on shift change from the mines. There is a third way out: on medevac to Tromsø, though this is just as likely to be via bodybag.

At the hotel bar, daytrippers drink imported vodka, watch World Cup playoffs on a Cold War television. Snap photos of Lenin's head floating on its plinth above the square. Leave within the hour, the tour boat gunning its motor into the coal-dusted harbour.

2

A tenth of the residents are children, the young people pulled into the mines or into service to keep the community running. Bar, hotel, restaurant, swimming pool—a slow modernization at the top of the world to appeal to tourists.

Once, the town was a place of colour and light, painted tin pushing back eight months of polar winter. Now, Cold War murals flake from brickwork; painted metal flowers droop in former gardens. Light standards line the square, bulbs shattered to time, to drunk-cast stones. It used to be home, of a sort. 2,500 in this port town at its height, the airliners coming and going weekly to the mainland. A secondary Russia, halfway to the North Pole from Murmansk.

Now, punched-out lights in an abandoned square, gardens long gone to frost heave and seed. Nobody wants to see it more clearly. Those who do cannot bear to stay.

3

The woman at the bar in the tourist hotel, pouring vodka few men can afford to drink on mine wages. Or the reed of a girl who wins the Miss North Pole pageant (when it runs), uncanny in stilettos.

This is a hard place for women, for families. Most are back in the Ukraine; the few here are freighted by dark: the winter, the cold. The mine and what it does to their men, lurking upslope from the harbour, tainting every fall of snow with coal dust.

They look out for each other, the women who stay. Mind children on shift days, share bread and soup when someone's husband has come in too late at night, drunk on cheap liquor, blown credit. They find small paths to beauty in the colour of a dress, a half-used tin of lavender paint for a child's bedroom. Prop each other up when the mine shudders into itself below sea level, flattens a man under ten tons of coal seam come thundering down the passageways like judgement.

4

The men dig for coal at the northernmost rim of the world, five hundred feet below the floor of the Barents Sea.

Arktikugol is headquartered in Russia, sidelined by Norway. When the seams claim a man, the mine closes just long enough for the body to be flown out, the tunnel cleared. Everyone is there to make money, a two-year stint, perhaps four, and then life can begin again somewhere else. Somewhere the cold doesn't harry their bones, the dark of midwinter challenging even the pitch shafts of the mine. The Arctic Coal Trust bleeds a trail of rubles across the sea to Svalbard so that Russia can keep a foothold on Spitsbergen now that Pyramiden has gone.

Above the miners' heads, their children learn the Russian words for birch tree, waterfall. For years, none of them will see a tree taller than polar willow, fuzz of leaves across the outcrops in spring. When they board the plane for Donetsk at the end of their fathers' shifts, they will feel they are shucking off all the darknesses of their short lives. As though the mine years will unravel on the ice below like a coal seam bisecting the Barents Sea.

Cusp

Of course I'm a teller
of mundane lies, such as: I'll try
never to lie to you. Such as:
the day after today the earth will
tilt on its axis towards the sun
again, the light will turn stronger,
it will be spring and you'll
be happy.

—MARGARET ATWOOD

1

Imagine this place coming out of the dark.

One of the guides, Freya, tells me of the first red crescent of sun reaching over the mountains. After a night of dancing, everyone drinking more than they could afford, how her limbs ran loose, guywiring toward summer. Winter draining from her body in one night, its long blue shadows pulling back into the crimp of the glaciers.

She tells me of the friends who didn't make it through the winter, gifting their names to me as though I can hold them with more tenderness, coming from away. There is no place for them on Spitsbergen; their bodies belong anywhere beyond those rocks and inlets. Why not Canada. I speak their names aloud, holding the sounds on my tongue.

My friend doesn't winter on the island. She stays through the long indigo nights as October devours the sun, takes her dog for a last run by lantern light, skirting the toe of the moraine. Her father's rifle rings against her shoulderblade like winter as she sights the dark for bears. In the morning, she will be gone, her partner waking to cool sheets, the husky's abandoned chain at the back of the house. Seasonally, she goes; he knows this. It is as expected as the snow. Each autumn, he feels the silence settle into his bones as the first storms cross the harbour.

2

Later, from a winter home in Stockholm or Catalonia, Freya will teach me her home country. Dwarf birch and sandwort, *fjellsyre*. Mountain crowberry, how the old women sought its dark fruit when they were young and pregnant, afraid to carry their babies into the Svalbard winter. Before the law that banished birth from the ice.

Under a sun that has vanished from Spitsbergen, she will sit out the afternoons and tell me of oysterplant, silver and blue in the gravel by the tide line, and how the mineral flats outside Longyearbyen gloss each summer with cottongrass plumes. Her home will have vanished into the dark, but I'll hear July in her voice: *kantlyng*, white arctic bell-heather, each blossom clenched around a drop of nectar. Where it grows close to the houses, children seek the small bells as avidly as bees, brief syrup of Arctic summer on their tongues.

3

My November birthday, and she'll send me a note over morning tea,
neat list of words unfolding itself like a map.

Noen blomster for deg:
Snøstjerneblom. Tundraarve. Polarblindurt. Dvergsoleie. Høgfjellkarse. Bleikrublom. Stivsildre.

Some flowers for you:
Tundra chickweed. Arctic mouse-ear. Polar campion. Pygmy buttercup. Alpine cress. Pale Whitlow-grass. Hawkweed saxifrage.

Across the world, a room blossoms.

4

Her friends know the mines, or, like her partner, the trawlers' decks.

If they are lucky, they come clear of the earth at poppy time, when the hills around the archipelago erupt in white and yellow for a few short days. The fortunate ones have someone at home who fingers slips of green into a glass on the supper table, brings the tundra's incandescent sweetness into a plain-framed room.

She carries cottongrass wands when she leaves the island. In the stillness that remains, her partner drinks coffee from her chipped blue mug, turns on lights one after the other in empty rooms. He will work the shipyards through the winter, day after day of rivet guns and steel wool, the sharp scent of pine. Count time by the first talon of sun hawking its way over the mountains, February light bringing her home.

Postcard from Svalbard

The soul should always stand ajar.

—EMILY DICKINSON

1

I can only imagine what they felt, those first explorers. So many humans in such compressed space, the reek of skin and barrelled meat. I come to this place hundreds of years later, barren as the stones in the face of summer, a woman setting her own markers in a history of men.

I watch twenty colleagues cycle through the same homesickness, days apart. Someone loving, another grieving. We have signed our lives to this voyage, this ship. There are days when it strikes us hard.

There are days, too, when personalities rub like pack ice, perpetual grind. We have learned quick exits from crowded rooms, how looks mean nothing but wrong place, wrong time. *Antigua* offers us little respite; there is no space on this ship to take anything personally.

2

I wish you could have seen the belugas breaching in Ny-Ålesund harbour, brought up by the long, light night. Today I learned what it means to live in sight of history, the thrum of it. The *Italia*'s mast hummed to such a pitch in the wind outside of town that we couldn't attune to a bear's approach. There are no edges here: none save the guards carry guns, and the sole markers of bear country are the signs telling us where the boundary begins. Everyone leaves their doors unlocked, and it is perfectly possible to have a stranger crash into your living room in the brilliant sun of two A.M., angry bear in tow.

There are days when the air is so clear, you can see for miles along the shore and not realize how far you've come. I did the same thing, too, when I stepped aboard the Boeing to leave home. I didn't know at the time how far this journey would take me, or how deep.

3

It's whispered there was a man who couldn't stand the closeness of the ship, the bright shiv of summer. He put ashore in Ny-Ålesund and begged at the research stations until his home country agreed to fly him back again. Just like that, one fewer body, one absent cycle of missing and longing. They tell us at the whaling stations how so many men vanished from the explorers' ships—to lead or scurvy, not circumpolar rescue.

I catch myself wondering what it would be like to flee for home. Even in summer, these slopes are picked out in ice. I imagine lifting into the sky and turning back to the long green days and the short, welcome nights. But there's no strength in me to beg a rescue. Daily, *Antigua* and I battle, seasickness and confinement and fear. But daily, she offers me gifts. Today, I hauled rigging, palms rope-burnt and raw, pleated out her mainsail like a wing. We're uneasy partners, this boat and I. She's not finished with me yet.

4

Each day, I heft the weight of my fear. On the drift ice, out with the Zodiac and lifejackets, one slip beneath the surface will steal your breath. On the beach, we've all developed a wariness, attuned every moment to the polar bear guards, the direction of their rifles. On board the ship, there seems scarcely enough room for me to circle inside my own skin, let alone dance with all these others.

The sound of kittiwakes mewling out over the water when I wake into the early morning. The worn planks of whaling stations, inscrutable and grey, and the sudden figure of a ranger rising from a bench by the door to stand and wave. The skim of the *Antigua* under sail, her nine-foot draft gifting us narrow channels, seaways closed to the larger ships. How we are, all of us, split open to this place at the top of the world.

I heft the weight of awe.

At the Face

Thousands of tired, nerve-shaken, over-civilized people are beginning to find out that going to the mountains is going home, that wildness is a necessity.

—JOHN MUIR

1

To watch a glacier calve is to watch time run in both directions at once.

The grey face is the old ice, pitted with history. The blue face is the fresh ice, brilliant and unscarred, razor-edged and untouched.

On deck, I watch the berg slew, wind-struck, swing its keen blue face away from land. Calving is a funny thing, birth and death gnarled inseparably. The new berg is a small knell for the glacier cracking its spine along the strait. Each summer a little thinner, ice bruised by a malignant swell of mountain coming clear beneath. In a scant decade, there will be only rock; the swaybacked ridge of the glacier will be as much a memory as the whale pods darkening the inlets along Svalbard's flank.

Birth and death, companion sides of the card. Up from Port Longyear, the Svalbard Seed Vault blunts its snout from the permafrost, stuffed with India, the ripe heart of the tropics. Mango and quince, rattlesnake bean, fennel. In this doomsday refrigerator, we trust our luck and double down. Millions of seeds packed into the frigid heart of the hill, waiting for that indistinct time after the glaciers have gone, shed their pockmarked faces into the fjord. Whatever that world resembles, we imagine rosemary and persimmon, white pine weeping with gum. Our vault a hedged bet, a long con. We are in this for the win.

Past *Antigua*'s hull, calved bergs float Janus-faced, wind bellying them out into the bay, crooked grey shadows back-cast. The blue crags of their birthplace already sun-slagging, gleam weathering to runoff sheen as we watch.

2

The old ice on these mountains carries centuries, the guides tell us.

When the others tire and go below deck, I linger at the railing, imagining Mount St. Helens, Bikini Atoll, and Nagasaki, the gunmetal glide of uranium from Canada's north. What we are capable of when we hollow the earth, small gods with hammer and tongs, refineries and their atmospheric smear. In our hands, Vulcan's desire to fashion in his own image, a cacophony of voices forming under his fingers. A face pinched from the clay, distant muttering from a tight-jointed box. Bright-eyed Pandora and her cargo of assassins.

Somewhere in this ice, Dachau plumes dark against the blue; London burns from a shop on Pudding Lane. I picture the ice sharding and Vesuvius issuing forth in a charred waft, the withering harmattan over the water from Marrakesh. A caftan of brittle wings.

3

New ice glares blue through the Svalbard dusk, the never-setting sun.

What will our record be, this clean face splitting, shape-shifting in looking-glass slides? Ragged Syrian fields driving the farmers to Damascus, Cascadia popping her knuckles along Qwe'Qwa'Sot'Em clam gardens. Soot plumes from the Paragominas as the Amazon burns; arsenic flares outside Accra where the Western computers burn.

What will this face remember? A Bastille Day truck attack in France. The Pulse Nightclub in Orlando. Dust and drought in the camps at Kigeme. Perhaps it will see the future rattling by like a deck of cards, sharp shuffle, dab hand. Ice thinning over a rictus of rock.

Threads

Why should we tolerate a diet of weak poisons, a home in insipid surroundings, a circle of acquaintances who are not quite our enemies, the noise of motors with just enough relief to prevent insanity? Who would want to live in a world which is just not quite fatal?

—PAUL SHEPARD

1

PCBS in Billesfjorden, Pyramiden leaking its slow veins out to sea.

Diesel leakage at Kapp Linné, the old Isfjord Radio building. Unable to contain due to polar winter.

Oil pollution from the storage ships wrecked at Kapp Laila.

Spring, and the aromatic hydrocarbon count in Longyearbyen as high as Zurich. Snowmobiles the culprit.

2

Four square kilometres of tailings into the Cariboo at Mount Polley.

380,000 litres of condensate into grizzly habitat at Grand Cache.

4.5 million litres of crude into the muskeg by Little Buffalo.

Half a million litres of sour crude into the Red Deer River.

230,000 litres of crude onsite at Elk Point.

60,000 litres of crude and hydrogen sulphide gas into the muskeg at
Red Earth Creek.

Regulators say, wildlife not affected.

3

Two crude oil spills a day in my home province, every day since 1975.

But here's what's missing: interprovincial pipelines. International pipelines. Anything considered a minor incident. *Read, 2,000 litres of crude into the muskeg.*

Anything that doesn't directly impact homes, economies. Anything not immediately seen.

Read, John O'Connor. Fort Chipewyan.

Everyone in Alberta knows what it means to live downstream.

4

So much of what impacts Svalbard comes from away, waterborne.
So much of what breaks my home comes up from beneath.

Land mascotted by something elemental. Bears front news feeds, harried and collared, bullied by cameras. Reduced to swimming for their lives, to grease bins and dumpster diving. Cubs born spoiled at the wrong end of a rifle.

The only wilderness we countenance: a safe remove.

Leaving Days

The dark is defeated. We have run out
of dark. We are darkness-deprived.
We rise from the runway, imagining night.

—ROBERT KROETSCH

1

The cemetery and I keep company into the long, light, final night.

I imagine all the ways of leaving this place: if you give birth or die, if you cannot hold a job or fall ill. If you cannot withstand the long dark. Svalbard poppies at my feet, outward creep of roots in ground that has refused caskets for seventy years because of the permafrost, its habit of stopping time.

I wonder whether this piece of ground, too, will change as the face of the Arctic changes; whether human life will be allowed to end its circle again here one day. What that will mean for the town, stilt-legged beside the harbour, if the permafrost turns to gruel. Yellow petals at my fingers, the sunlight shifting to gauze at two A.M. *Antigua* riding the dockside swells at the edge of sight, tied in for resupplying before she skims out again, carrying a different set of faces.

2

We carry this space with us when we go.

In photographs and recordings, etchings and climate data, we carry the feel of a Svalbard summer. On the sun-brittled carapaces of prairie sailboats, I'll find my feet and bear down; I'll cast back to *Antigua*'s decks, the come-from-away skeletons of the whaling stations under endless noonhour sun.

The way of all expeditions: at the end, a slow fragmenting, everyone compassed toward home. Pyongyang and Halifax, Seattle and Sacramento, London, the Alberta bush. We are flotsam, travelling greedy tides back to our own small spaces of dark. Our own welcome nights.

3

When we wake, *Antigua* has re-crewed and gone, luffing quietly out from the harbour in the early morning. I imagine the crew watching for open water before hauling up her mainsail, feel the familiar creosote burn of rope across my palms. Two weeks of learning that ship, every pace of her deck, sound of wind in the rigging.

Nothing left at her berth but water tossing greyly along the shingle.

4

Thirty-six hours home to a patch of boreal on the other side of the world. I gather darkness as I go: scant couple of hours in Tromsø, a handful in Halifax, where I wait through muggy midsummer dark for an early-morning transfer west.

In Toronto, I see my first skyscrapers in over a month, slicing through heat haze above the lake. My mind on glacier ice, its million refractions, its cannonball weight in the palm, all air bubbles gone.

Thirty-six hours before the plane curves out of the sky to Edmonton and the boreal rises north like night. A cool swoop. A raven's wing.

Song to the Boreal

Words act as compass; place-speech serves literally to en-chant the land—to sing it back into being, and to sing one's being back into it.

—ROBERT MACFARLANE

1

I return home craving the forest.

Too late for spruce-tip tea, branch candles gone long and waxy green,
I satisfy myself with fireweed jelly, early saskatoons, ground raspberry.
Browse the canes and brambles, startling deer, white-crowned sparrows.
I'm looking for berries, pulp and dark juice. Things that leave a mark.

2

Svalbard has scoured me clean, beach glass at a king tide. I reek of salt, the taste of it ground into my lips, my skin burnt by wind. How will you know me like this, soil scuffed from my palm lines? How will I know myself?

I come back marvelling at trees, unconvinced by dwarf birch, polar willow. Black spruce, trembling aspen, wildling crab; I pull green apples, sink my teeth for the astringency on my tongue.

Wild harvest means coming home: I pluck chickweed, lamb's quarters, dandelion greens, splash vinegar to brighten, brew bush tea. I am hungry at my roots for the forest, lilt of honey on my tongue a kind of grounding.

3

Sing diamond willow for carving, red elder bitter with cyanide. Wild gooseberry and twinflower, bog orchid, sedge.

Pluck burdock, plantain, styptic yarrow.

Give me early afternoon under the canopy, bush tea rumbling with bees; let me sink glassine fingers into the moss and I'll know I'm home.

4

Mid June, and the boreal kindles like tallow.

No toxins across the water; here, one Players on the frontage grass and everything goes up, black spruce rickety with gum.

Each summer a little longer, a little hotter, and still the quads idling out in the muskeg, stink of gas in the deep woods. Used to be fork lightning, now it's Friday night bush parties, chainsaw sparks. I come home from water and ice to a forest spitting resin like midsummer rain. We watch the news, the wind. There's not enough water in Svalbard to put this forest out if it burns.

5

Peat fires know to go underground, run out along the roots. A whole stand burning slowly from beneath.

Our first winter on the land, we lit a bonfire in a space of cleared ground to keep warm. Diamond willow, balsam poplar; everything we felled went into the flames, a long, slow burn. Potatoes in the coals and they would roast all day, pried from charred skins at twilight as the deep cold came down.

We learned that fire in the north is subterrain, that even the snow refused our circle, its heap of ash. Dug out the ground a week later and it was hot enough that you couldn't hold the soil without gloves. It smelt of roots and hair, a living, feral scent. Those are the fires you learn to fear, the ground going out from under your feet in an instant. Nothing left but cinders grasshoppering into the dark.

6

Bedstraw, marsh marigold, bunchberry. White conks on diamond willow that burn with a slow smoke, scent of anise seed. Take the pain out of bad lungs.

This forest is home in its ways of knowing, being known. Grouse drumming in the deep woods, ropes of toad spawn in the pond. Underfoot, wild strawberry, wood violet, porcupine spoor. Move with a measured pace and catch the boreal staring back.

7

Apricot jelly.
Bear's head.
Black.
Comb tooth.
Common puffball.
Fairy ring.
Horse.
Ink cap.
Larch bolete.
Lobster.
Meadow.
Mica cap.
Oyster.
Rosy gomphidius.
Saffron milk cap.
Scaly hedgehog.
Shaggy mane.
Slimy spike cap.
Western giant puffball.
Wood ear.
Yellow swamp russula.
After a burn, polyphony.

8

Late June, and the boreal is alight.

Half a world away, glaciers cleave from their footholds like jumped tracks, their faces reconfigured in centigrade increments.

I come home to the forest with a vastness behind my eyes, a consciousness of rock and light, an awareness of what's lost without the dark. The small brown flicker of a wren in the amaranth, the long white wing of a gull. In the back of my mind, *Antigua* ships sail, swings toward her mooring. A green heel of forest, a vanishing rime of frost.

Summer opens its palm.

Notes

Epigraph from *The Black Grizzly of Whiskey Creek* by Sid Marty, pg. 24, Emblem, 2008.

Lines Toward Ice epigraph from "Landscape With The Fall of Icarus'" by William Carlos Williams, from THE COLLECTED POEMS: VOLUME II, 1939–1962, copyright ©1962 by William Carlos Williams. Reprinted by permission of New Directions Publishing Corp.

Pyramiden epigraph from *Landmarks* by Robert Macfarlane, pg. 25, Hamish Hamilton, 2015.

Ornithomancy epigraph from *Frissure: Prose Poems and Artworks*, Polygon, 2013 (imprint of Birlinn Limited) by Kathleen Jamie (Brigid Collins illustrator).

Night epigraph from "Garden (sept unit)" by Monty Reid, originally published in *Garden*, pg. 12, Chaudiere Books, 2014.

Bone epigraph from "Garden (nov unit)" by by Monty Reid, originally published in *Garden*, pg. 28, Chaudiere Books, 2014.

The Men at the Edge of the World epigraph by Jalaluddin Rumi, published in *Rumi: The Book of Love*, translated by Coleman Barks, pg. 124, HarperOne, 2003.

She Becomes the Ocean epigraph from "Mad Girl's Love Song" by Sylvia Plath, originally published in *Mademoiselle*, pg. 358, August 1953, New York, USA, Condé Nast Publications Ltd.

Afloat epigraph from "Lightenings" by Seamus Heaney, published in *Seeing Things*, pg. 56, Farrar, Straus and Giroux, 1993.

Barenstburg epigraph from "The New Faces" by William Butler Yeats, published in *W.B. Yeats: Selected Poetry*, pg. 147, Penguin, 1991.

Cusp epigraph from "Small Poems for the Winter Solstice" by Margaret Atwood, originally published in *The Missouri Review*, Issue 5.2 (Winter 1981–82).

Postcard from Svalbard epigraph from "The soul should always stand ajar" by Emily Dickinson, published in *The Complete Poems of Emily Dickinson*, Little, Brown, 1924. Bartleby.com, 2000. www.bartleby.com/113/. [18 June 2018].

At the Face epigraph from *Our National Parks* by John Muir, pg. 11, Houghton Mifflin, 1901.

Threads epigraph from Paul Shepard, quoted in *Silent Spring* by Rachel Carson, pg. 12, Mariner Books, 2002. Originally published in 1962 by Houghton Mifflin.

The statistics mentioned in Threads are pulled from the main page and oil leak lookup link in the Global News article "Crude Awakening: 37 Years of Oil Spills in Alberta." http://globalnews.ca/news/571494/introduction-37-years-of-oil-spills-in-alberta/

Leaving Days epigraph from "Driving to the Airport at Five AM" by Robert Kroetsch, originally published in *Too Bad: Sketches Toward a Self-Portrait*, pg. 35, University of Alberta Press, 2010.

Song to the Boreal epigraph from *Landmarks* by Robert Macfarlane, pg. 22, Hamish Hamilton, 2015.

Acknowledgements

I'M GRATEFUL TO so many for their support during my trip to the Arctic and the subsequent slow gestation of this collection.

First, to the wonderful people who made my journey to Svalbard possible: Dr. David Atkinson, former President of MacEwan University in Edmonton, for wholeheartedly supporting my travels halfway around the world; and Aaron Turner of the Arctic Circle Expeditions for organizing and herding a diverse collection of writers, artists, and scientists safely across Svalbard.

To my madcap travelling companions onboard the *Antigua*, thanks for good friendship, 24-hour-daylight adventures in one of the most beautiful places on earth, and a plethora of stories to recount back home.

Thanks to Mary Pinkoski, the 2014 Poet Laureate of Edmonton, for the invitation to read from the early pieces at the CBC Centre Stage; Elizabeth Withey, former book reviewer at the *Edmonton Journal*, for a supportive write-up on my journey; Andrew Forster and Lindsey Holland of *The Compass* (UK) for their thoughtful interview on my Arctic travels and off-grid life at home in Canada, and for publishing sample pieces from the collection; Patricia Tempest, Visual Display Designer at MacEwan University Bookstore, for outfitting me for an Arctic expedition; and Carmen Rohac at the University of Alberta for publishing a snapshot article on my journey north. To my editor at the University of Alberta Press, Peter Midgley, my gratitude as ever for that uncanny ability to see both the complete arc of a book and the necessary particulars, and for many years of friendship. To the publishing team at the University of Alberta Press (Linda Cameron, Douglas Hildebrand, Mary Lou Roy, Duncan Turner, Cathie Crooks, Monika Igali, Alan Brownoff, and Basia Kowal), thanks and admiration. To Carlen Lavigne, for thoughtful and intelligent feedback, and to Torben Andersen, for ongoing literary support, much appreciation.

Thanks to the marvellous Hamish Robinson at Hawthornden Castle in Scotland for looking after the Spring 2017 group so well, and for being the most engaging host one could hope to meet; to Mary, Georgie, and Ruth for their warm-hearted care of our gaggle of writers while we worked on our various book projects; and to the admissions board and the generosity of the late Drue Heinz, DBE, for granting us the gift of four weeks as Fellows in such an enchanted place.

As ever, I am grateful to the following institutions for their unwavering support through the years of this project: Red Deer College; MacEwan University (the Research, Scholarly, and Creative Activities Fund); MacEwan University Bookstore; the Alberta Foundation for the Arts; the Edmonton Arts Council; and CBC Centre Stage/CBC Radio Edmonton.

OTHER TITLES FROM THE UNIVERSITY OF ALBERTA PRESS

Wells

JENNA BUTLER

Through poetry, Jenna Butler pieces together the life of a cherished grandmother lost to Alzheimer's.

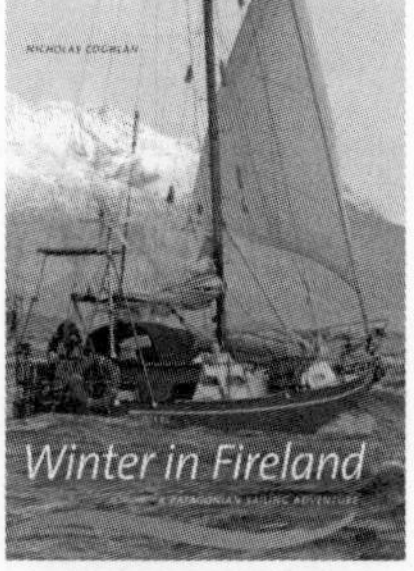

Winter in Fireland

A Patagonian Sailing Adventure

NICHOLAS COGHLAN

British-Canadian diplomat and wife sail from Cape Town to Cape Horn in their 27-foot boat.

Rain Shadow

NICHOLAS BRADLEY

Poetry that explores our fraught relationship with nature: playful, serious, heartsore.

More information at www.uap.ualberta.ca